Tarek Saidani
Mohammed RASHEED
Habiba Aity

ELECTROSTATIC

Tarek Saidani
Mohammed RASHEED
Habiba Aity

ELECTROSTATIC

ScienciaScripts

Imprint

Cover image: www.ingimage.com

This book is a translation from the original published under ISBN 978-620-6-70410-2.

Publisher:
Sciencia Scripts
is a trademark of
Dodo Books Indian Ocean Ltd. and OmniScriptum S.R.L publishing group

120 High Road, East Finchley, London, N2 9ED, United Kingdom
Str. Armeneasca 28/1, office 1, Chisinau MD-2012, Republic of Moldova, Europe
Managing Directors: Ieva Konstantinova, Victoria Ursu
info@omniscriptum.com

Printed at: see last page
ISBN: 978-620-8-59843-3

Table of contents

Foreword

This booklet is intended for students in the first year of the LMD system, Science and Technology (ST) and Material Sciences (SM) fields, second semester of the academic year

This book includes a reminder of the electrostatics course, exercises with answers, and brings together the fundamental concepts of electrostatics, in which we study the concepts, phenomena and laws reserved for the electricity of locally "immobile" charges.

The course review concludes with a series of well-chosen exercises, followed by detailed answers, to help users of this document fully grasp the concepts introduced in the section containing the course review.

In fact, we hope that this booklet will provide our students with a good grounding and pave the way for the future.

Finally, we welcome all comments and suggestions from users, both teachers and students.

1. Electrostatic charges and fields

1.1 Electrification

Electrification is the production of electricity through interaction between two electric charges (particles) located in space.

These particles fall into two categories: insulators or dielectrics (glass, nylon, plastic, ebonite, etc.) and conductors (earth, human body, metals, water).

There are two types of electrification, positive and negative. A body that is not electrified (charged) is said to be neutral [1].

- When two bodies are of the same nature, they repel each other. This is known as repulsion (Figure 1).
- When the two bodies are different in nature, they attract each other. Attraction occurs (Figure 1).

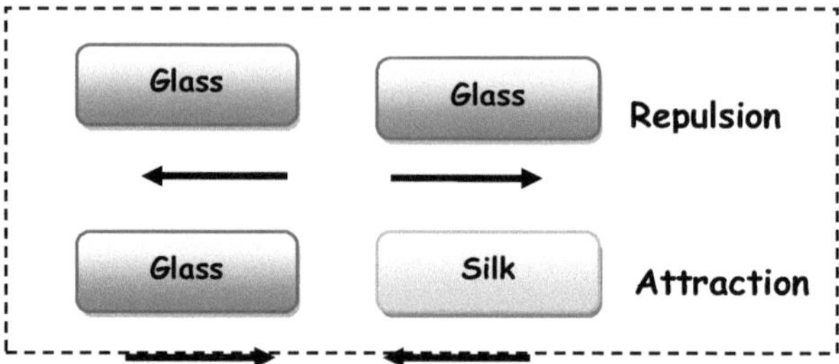

Figure 1: Electrification

1.2 Means of electrification

- Electrification by friction : Electrons are stripped from the neutral body being rubbed and then transferred to the second body, which becomes positively electrified (glass, silk, wood, rabbit hair, cat hair, mica, sulphur, amber, resin, ebonite, celluloid) [1].
- Electrification by contact with an already electrified body: A body electrified by friction (glass) is brought close to another neutral body (polystyrene ball surrounded by a conductive material) until contact is made. Through interaction, the two bodies (the glass and the ball) will be charged with electricity of the same sign and will repel each other. The result is repulsion [1].

- Electrification by influence: The electrified body (glass) is brought close to the second initially neutral body (the ball), without touching it. The second body (ball) is attracted to the first body (glass) by the influence effect.
- Generator electrification: An electrical generator has positive charges on one terminal and negative charges on the other. If one of the terminals is connected to an initially neutral body using a metal wire, the body becomes electrified.

2. The electrostatic force

2.1. Coulomb's law

Consider two identical point chargesq_1 andq_2 separated by a distance "r". Each of the charges exerts on the other a proportional electrostatic interaction force proposed by Coulomb's law by analogy with Newton's law of universal gravitation (figure I.2) [1].

$$\overrightarrow{F_{1/2}} = k\,\frac{q_1 q_2}{r^2}\,\overrightarrow{U_{1/2}}$$

With $k = \frac{1}{4\pi\varepsilon_0} = 9.10^9\ Nm/c^2$

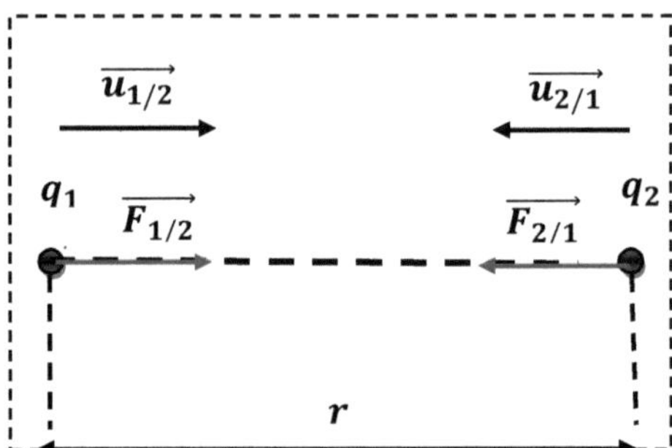

Figure 2: Vector representation of the electrostatic force

2.1.1 Properties of force

- One module : $K\frac{|q_1|.|q_2|}{r^2}$
- One direction: a straight line passing through the two loads.
- Meaning :
  - $+q_1q_2 > 0$ (repulsion) (see Figure I.1)
  - q_1 -. $+q_2 < 0$ (attraction)
  - q_1 -. $-q_2 > 0$ (repulsion)
- Principle of action and reaction : $\overrightarrow{F_{1/_2}} = -\overrightarrow{F_{2/_1}}$
- The charge on the electron: $q_{e^-} = -1.6 .10^{-19}$ C

2.2 fields

The electric field is defined by the region of space where any particle is subject to the action of an electric force [2].

- Case of a particle of mass m: it is subject to the force of gravity (its weight)$\vec{\boldsymbol{p}} = \boldsymbol{m}.\vec{\boldsymbol{g}}$

- Case of a particle with charge q: it is subjected to the electric force : $\vec{F} = q.\vec{E}$ with $\vec{E} = k.\frac{q_1}{r^2}.\vec{u}$ (The field created by the charge q at a distance r follows the direction of $\vec{u}$).

2.2.1. Electric field created by a point charge

A point electric charge at point O creates at point M at a distance r the electric field vector $\vec{E}$. The expression $\vec{E}$ is given by the following formula:

$$\vec{E}(M) = k\frac{q}{r^2}\vec{u}$$

The electric field goes from positive charge to negative charge or to infinity.

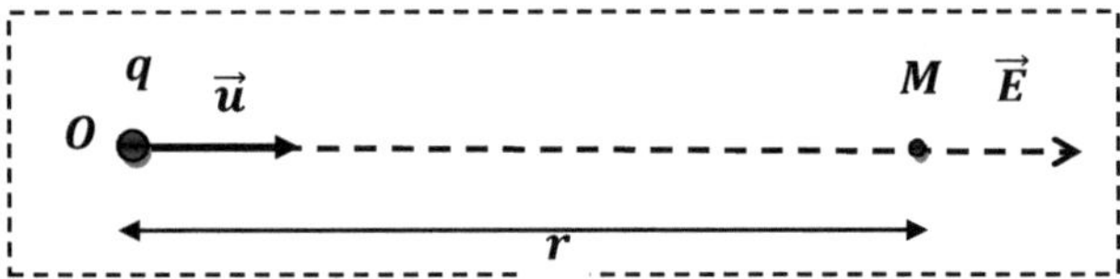

Figure 3: Vector representation of the electrostatic field

2.2.2. Electric field created by a set of point charges (superposition principle)

Consider two point charges q_1 and q_2 at distances r_1 and r_2 as shown in Figure 4.

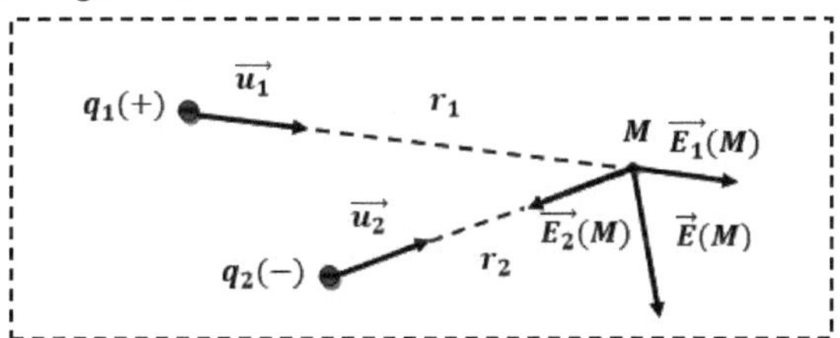

Figure 4. Superposition principle

The field $\vec{E}$ created at a point M by the two point charges is the sum of the two fields $\overrightarrow{E_1}$ (M) and $\overrightarrow{E_2}(M)$ created by each of the charges q_1 and q_2 . The expression $\vec{E}(M)$ is given by the following formula:

$$\vec{E}(M) = \overrightarrow{E_1}(M) + \overrightarrow{E_2}(M) = k\frac{q_1}{r_1^2}\overrightarrow{u_1} + k\frac{q_2}{r_2^2}\overrightarrow{u_2}$$

Case of n point charges , ,$q_1 q_2 q_3$, ... q_n :

$$\vec{E} = \sum_{i=1}^{n} k\frac{q_i}{r_i^2}\overrightarrow{u_i}$$

2.2.3. Field created by a continuous distribution of charges

Let there be a set of charges q_i distributed over an element. Each charge dq located in this element creates an elementary field.

- On a wire of length dl with a linear density λ , (figure 5.a) dq = λ dl.

$$\vec{E} = \frac{1}{4\pi\varepsilon_0}\int \frac{\lambda\, dl}{r^2}\vec{u}$$

- On a charged surface with a surface density σ, (figure 5.b) dq= σ dS.

$$\vec{E} = \frac{1}{4\pi\varepsilon_0}\iint \frac{\sigma\, ds}{r^2}\vec{u}$$

- On a volume V loaded with a volume density ρ, (figure 5.c) dq = ρ dV.

$$\vec{E} = \frac{1}{4\pi\varepsilon_0}\iiint \frac{\rho\, dv}{r^2}\vec{u}$$

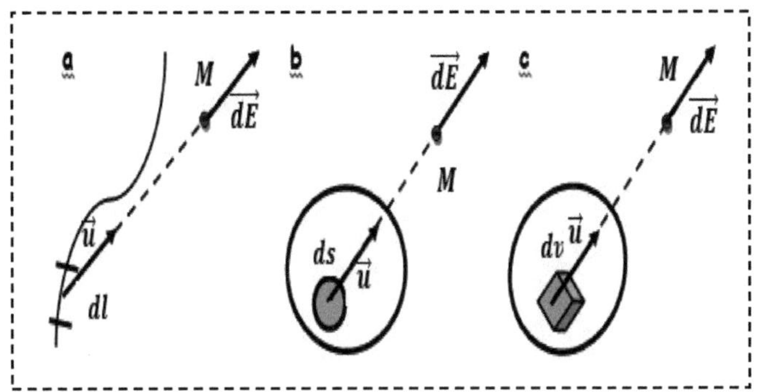

Figure 5: Continuous distribution

2.3. Electrical potential

A charge q, placed at a point M and creating an electric field vector$\vec{E}$ from its elementary displacement$\overline{dl}$, has a potential energy *Ep =q.V [2]*. The electrostatic field$\vec{E}$ is derived from a scalar potential : $\vec{E}.\overline{dl} = -dv$

The electric potential is written as :

$$v = \frac{1}{4\pi\varepsilon_0}.\frac{q}{r} + cte$$

- n point charges , ,$q_1q_2q_3$, ... q_n : $v = \sum_{i=1}^{n} k\frac{q_i}{r} + cte$
- continuous distribution of linearly distributed loads

$$v = k\int_l \lambda\frac{dl}{r} + cte$$

- continuous distribution of charges distributed over the

$$v = k \iint_s \sigma \frac{ds}{r} + cte$$

- continuous distribution of loads

$$v = k \iiint_v \rho \frac{dv}{r} + cte$$

2.4. From field to potential and from potential to field

At any point M (x, y, z) in space, there is a combination of two functions, vector (the field$\vec{E} = \vec{E}\,(x, y, z)$) and scalar V = $V\,(x, y, z)$.

It is given

$$-dv = \overline{\mathrm{E}}.\overline{\mathrm{dl}} = \mathrm{E_X}\,\mathrm{dx} + \mathrm{E_y}\,\mathrm{dy}\ + \mathrm{E_z}\,\mathrm{dz}$$

Knowing that

$$dv = \frac{\partial v}{\partial x}\,\partial x + \frac{\partial v}{\partial y}\,\partial y + \frac{\partial v}{\partial z}\,\partial z$$

By identification :

$$E_X = \frac{\partial v}{-\partial x}\,, E_y = \frac{\partial v}{-\partial y}\,, \qquad E_z = \frac{\partial v}{-\partial z}$$

We write :

$$\vec{\boldsymbol{E}} = -\overline{\boldsymbol{gradv}}$$

2.5. Field lines and equipotentials

Field (or force) lines are curves tangent to the field vector at each point (see figure 6): force tubes are sets of field lines that pass through all the points on a curve [3].

- The orientation of the lines follows that of the electric field.
- The field line is oriented from the highest to the lowest potential.
- The electric field is strongest where equipotentials are tightest.
- Equipotential lines and surfaces are the geometric locus of points of common potential.
- One of the properties of field lines is that they are perpendicular to equipotentials (Figure 7).

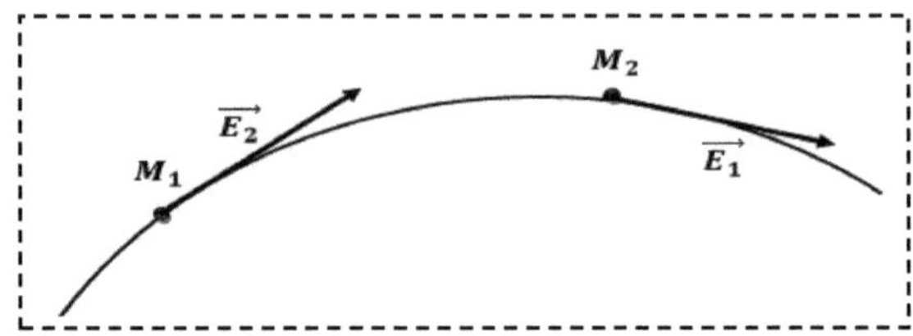

Figure 6. Electrostatic field line

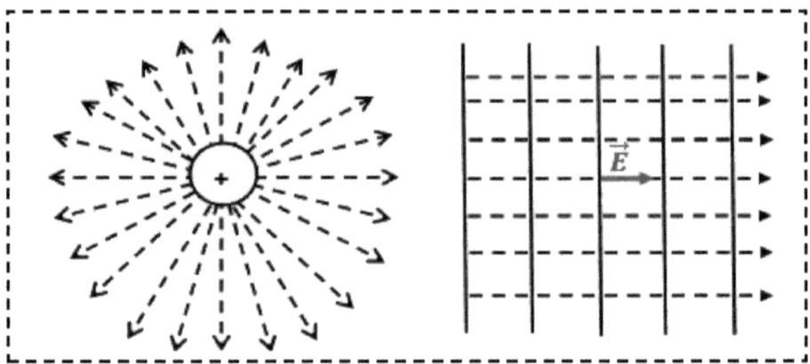

Figure 7. Directions of the electrostatic field lines

- If the charge Q is positive (+), the field is directed from the charge outwards (Figure I.7).
- If the charge Q is negative (-), the field is directed outwards towards the charge.

3. Work and Energy

3.1 Work of electrical force

An electric charge q is placed at a point in space where there is an electric field$\vec{E}$. It is then subject to the action of an electric force :

$$\vec{F} = q\,\vec{E}$$

The work done by this force during an elementary displacement$\overline{\mathrm{dl}}$ is :

$$dW = \overrightarrow{F}.\overrightarrow{dl}$$

During a journey AB, we have :

$$W = \int_{B}^{A} \vec{F}.\overrightarrow{dl}$$

Or :

$$\boldsymbol{W_B^A = q\,(V_A - V_B)}$$

3.2 Potential energy

The potential energy of a point charge placed in an external field is defined as the work of the electrostatic force acting on the charge, for a displacement of the charge from point M, where it is located and where the potential is V_M, to a reference point R, where the charge is no longer subject to the action of the external field. At this point, the potential is zero: $V_R=0$, Either:

$$E_p(M) = \int_M^R \vec{F}.\overrightarrow{dl} = q\int_M^R \vec{F}.\overrightarrow{dl} = q(V_M - V_r)$$

Therefore :

$$\boldsymbol{E_p(M) = qV_M}$$

Coulomb's force is therefore conservative and its work between any two points does not depend on the path taken. It is derived from a potential energy $E_p(M) = qV_M + cte$ and we write :

$$\boldsymbol{-dE_p(M) = \vec{F}.\overrightarrow{dl}}$$

3.3 Internal energy of a distribution of electrical charges

Let us consider two chargesq_1 andq_2 placed respectively at the pointsM_1 andM_2 distant from$M_1M_2= r_{12}$. In order to define the internal energy of the system of two charges, it is assumed that the chargeq_1 placed atM_1 is fixed and that the chargeq_2 approachesM_2 from an infinitely distant position.

The work that an operator would have to do to bring the loadq_2 up toM_2 without any variation in kinetic energy is given by:

$$W_{op} = \int_{\infty}^{M_2} \overrightarrow{F_{OP}}.\overrightarrow{dl} = -\int_{\infty}^{M_2} \overrightarrow{F_{1/2}}.\overrightarrow{dl} = q_2 \int_{\infty}^{M_2} \overrightarrow{E_1}.\overrightarrow{dl}$$

$\overrightarrow{F_{1/2}}$ is the electrostatic force exerted by the charge q_1 on the chargeq_2 . Let the internal energy of the system of two chargesq_1 andq_2 be the expression:

$$\boldsymbol{U = W_{op} = q_2.V_1 = K\frac{q_1q_2}{r_{12}}}$$

4. Electric dipole

The arrangement of two identical point charges of different signs forms an electric dipole (figure 8) [3]. A dipole is characterised by its electric dipole moment or electric moment:

$$\vec{P} = q.\vec{a}$$

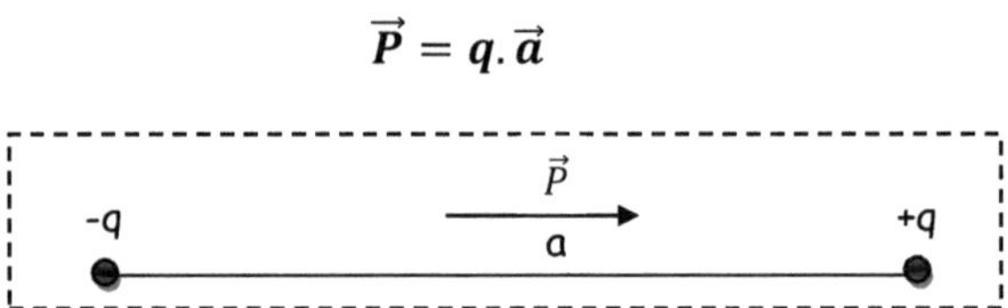

Figure 8. Electrical dipole moment

This arrangement appears in certain molecules such as: HCl,H_2O, CO, etc... as shown in the figure below.

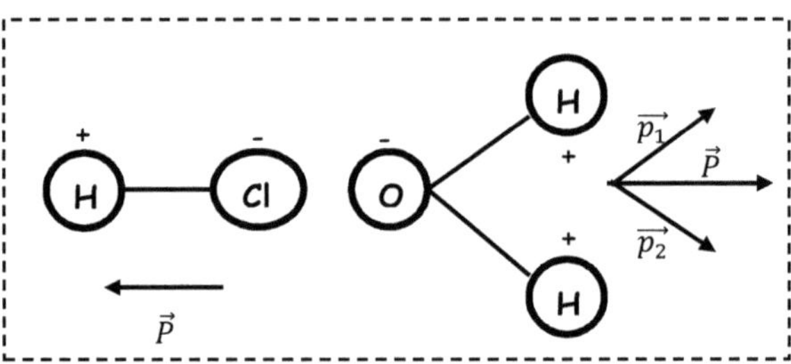

Figure 9. Electrical dipole moment

4.1. Potential created by a dipole

The potential at point M due to the dipole (Figure 10) is given by :

$$v = k\left(\frac{q}{r_1} - \frac{q}{r_2}\right) = k.q(\frac{r_2 - r_1}{r_1 r_2})$$

If the distance r is greater than a, then :

$$r_2 - r_1 = a.cos\theta \; et \; r_1 r_2 = r^2$$

And we'll have :

$$\boldsymbol{v = k.q.a\frac{cos\theta}{r_2} = k.p\frac{cos\theta}{r_2}}$$

4.2 Electric field created by a dipole

The relationship between field strength and potential is :

$$dv = -\vec{E}.\overrightarrow{dl}$$

The gradient of a function f is written as :

- In Cartesian coordinates : $\overrightarrow{grad}f = \frac{\partial f}{\partial x}\vec{\imath} + \frac{\partial f}{\partial y}\vec{\jmath} + \frac{\partial f}{\partial z}\vec{k}$
- In polar coordinates: $\overrightarrow{grad}f = \frac{\partial f}{\partial r}\overrightarrow{e_r} + \frac{1}{r}\frac{\partial f}{\partial \theta}\overrightarrow{e_\theta}$

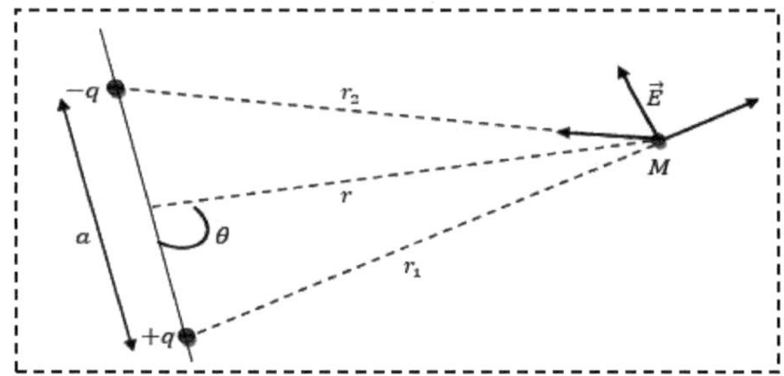

Figure 10. Electric field created by an electric dipole

With :

$$\vec{E}\begin{cases}E_r\\E_\theta\end{cases} et\ \overrightarrow{dl}\begin{cases}dl_r = d_r\\dl_\theta = r.d_\theta\end{cases}$$

$$dv = -\vec{E}.\overrightarrow{dl}$$

$$\Rightarrow\begin{cases}dv = -(E_r.d_r + E_\theta.d_\theta)\\ dv = \frac{\partial v}{\partial r}d_r + \frac{\partial v}{\partial_\theta}d_\theta\end{cases}$$

By determination, we obtain :

$$\vec{E}\begin{cases}E_r = -\frac{\partial v}{\partial r}\\ E_\theta = -\frac{1}{r}.\frac{\partial v}{\partial_\theta}\end{cases} \text{an } v = k.p\frac{cos\theta}{r_2}$$

$$\Rightarrow\begin{cases}\boldsymbol{E_r} = -\frac{\partial v}{\partial r} = \frac{2k.p.cos\theta}{r^3}\\ \boldsymbol{E_\theta} = -\frac{1}{r}.\frac{\partial v}{\partial_\theta} = \frac{k.p.sin\theta}{r^3}\end{cases}$$

5. Electric field flux: Gauss's theorem

5.1. Representation of a surface

A surface S is decomposed into small elements dS, represented as vectors and pointing in an arbitrary direction which is retained for all the elements of the surface S (Figure 11) [4].

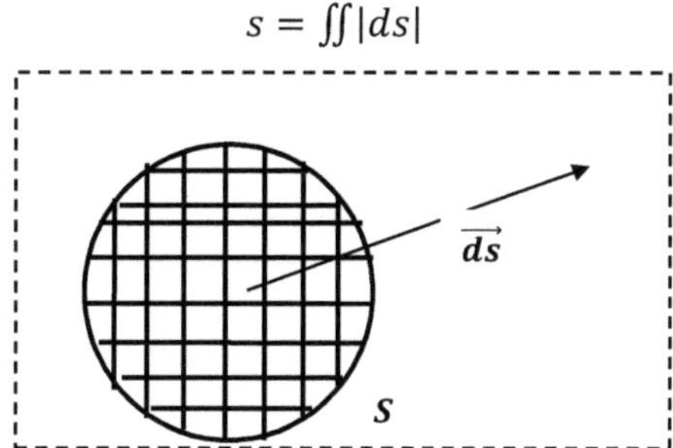

Figure 11. Representation of a surface

5.2. Flow of the electrostatic field vector across a surface

The elementary flux$d\Phi$ of the vector$\vec{E}$ across the surface dS is the scalar quantity defined as follows:

$$d\Phi = \vec{E}.\overrightarrow{ds} = E.ds.cos\theta$$

The overall flux through the surface S is obtained by integration:

$$\boldsymbol{\Phi} = \iint \vec{\boldsymbol{E}}.\overrightarrow{\boldsymbol{ds}}$$

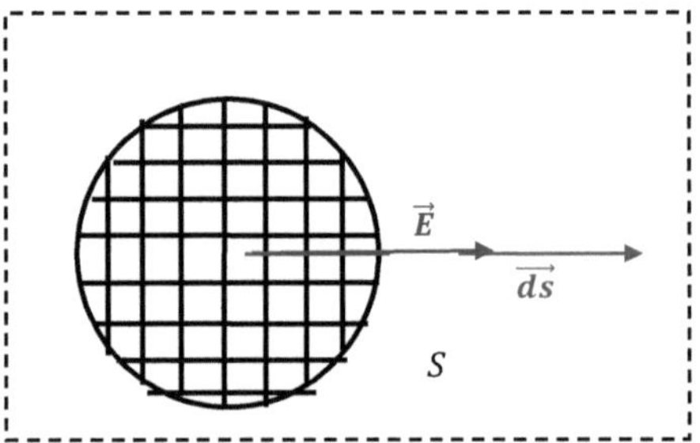

Figure 12. Flux$\vec{E}$ across surface

5.3. Excitation vector or electrical displacement

In electromagnetism, Maxwell introduced a vector $\vec{D}$ which he called electric displacement (electric excitation). This is linked to the electric field by the expression:

$$\vec{\boldsymbol{D}} = \boldsymbol{\varepsilon}.\vec{\boldsymbol{E}}$$

In the particular case of a vacuum, the previous expression becomes

$$\vec{D} = \varepsilon_0.\vec{E}$$

5.4. Gauss's theorem

The flux of the electric displacement vector through a closed surface S surrounding charges Qi is equal to the sum of these charges.

$$\oiint \vec{D}.\overrightarrow{ds} = \sum_{i=1}^{n} Q_i$$

In the special case of a vacuum, the electric field flow through a closed surface surrounding charges Qi is :

$$\oiint \vec{E}.\overrightarrow{ds} = \sum_{i=1}^{n} \frac{Q_i}{\varepsilon_0}$$

Exercise 1:

Silk is rubbed against a glass. During the process, 40.10^{12} electrons are transferred from the glass to the silk. Before the process, the silk and the glass are assumed to be neutral. What is the charge carried by the glass and the silk after the process? Data: the electrical charge of an electron is e = -1.602. C.10^{-19}

Exercise 1 solution:

During the friction, n electrons will pass from the glass to the silk, n= 40. .10^{12}

Before the process, the silk and glass are assumed to be neutral$q_v = q_s = 0\ C;$

Total load $Q_T = q_v + q_s = 0\ C$

After the process, the new silk and glass fillers are:q'_v and ;q'_s

Total load $Q'_T = q'_v + q'_s$

The system (glass+ silk) is insulated so the total charge is maintained:

$$Q'_T = Q_T \Rightarrow q'_v + q'_s = 0$$

$$\Rightarrow q'_s = -q'_v = n.e$$

$$\Rightarrow \boldsymbol{q'_s = n.e = -64.10^{-7}}$$

and

$$\boldsymbol{q'_v = -q'_s = 64.10^{-7}\ \mathrm{C}}$$

Exercise 2:

Two identical conducting spheres carry chargesq_1 andq_2 respectively. They are brought into contact and then separated. Determine the chargesq'_1 andq'_2 they take on, the direction of electron transfer and the number of charges transferred in the following cases:

1. $q_1 = +5.10^{-8}$ C and $Cq_2 = 0$
2. $q_1 = +4.10^{-8}$ C and $Cq_2 = +9.10^{-8}$
3. $q_1 = +2.10^{-8}$ C and $Cq_2 == -7.10^{-8}$

Exercise 2 solution:

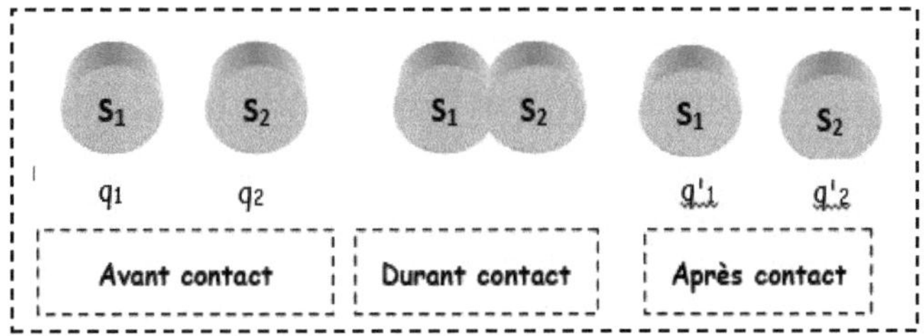

Figure 13

The system of two spheres is isolated, so the total charge is conserved:

$$Q_T = Q'_T \Rightarrow q_1 + q_2 = q'_1 + q'_2$$

The two conductive spheres are identical:

$$q'_1 = q'_2 = q' \Rightarrow q_1 + q_2 = 2q'$$

Hence :

$$q' = \frac{q_1 + q_2}{2}$$

The number of electrons transferred :

$$n = \frac{|\Delta q|}{|e|} = \frac{|q'_2 - q_2|}{|e|} = \frac{|q'_1 - q_1|}{|e|}$$

1. $q_1 = +5.10^{-8}$ **(C) and**$q_2 = 0$ **(C)**

$$q' = \frac{q_1 + q_2}{2} = \frac{5+0}{2}10^{-8} = 2,5.10^{-8}\ C$$

Electrons are transferred from theS_2 sphere to theS_1 sphere.

The number of electrons transferred:

$$n = \frac{|\Delta q|}{|e|} = \frac{|2,5-0|}{|-1,6.10^{-19}|}10^{-8} = 1,56.10^{11}$$

2. q1= $+4.10^{-8}$ **C and** q2 = $+9.10^{-8}$ **C**

$$q' = \frac{q_1 + q_2}{2} = \frac{4+9}{2}10^{-8} = 6,5.10^{-8}\ C$$

Electrons are transferred from theq_1 sphere to theS_2 sphere.

The number of electrons transferred :

$$n = \frac{|\Delta q|}{|e|} = \frac{|6,5-9|}{|-1,6.10^{-19}|}10^{-8} = 1,56.10^{11}$$

3. q1= $+2.10^{-8}$ **C and** q2 = -7.10^{-8} **C**

$$q' = \frac{q_1 + q_2}{2} = \frac{2-7}{2}10^{-8} = -2,5.10^{-8}\ C$$

Electrons are transferred from theS_2 sphere to theS_1 sphere.

The number of electrons transferred :

$$n = \frac{|\Delta q|}{|e|} = \frac{|-2,5 - 2|}{|-1,6.10^{-19}|} 10^{-8} = 2,81.10^{11}$$

Exercise 3:

The figure below shows a pendulum consisting of a wire of length l = 10 cm and a ball of mass m = 9 g carrying an electric chargeQ_1 = +2.10^{-8} C. A point chargeQ_2 = -5 is placed at a distance d = 4 cm from the ball. 10^{-8} C. Let g = 10m/ .s^2

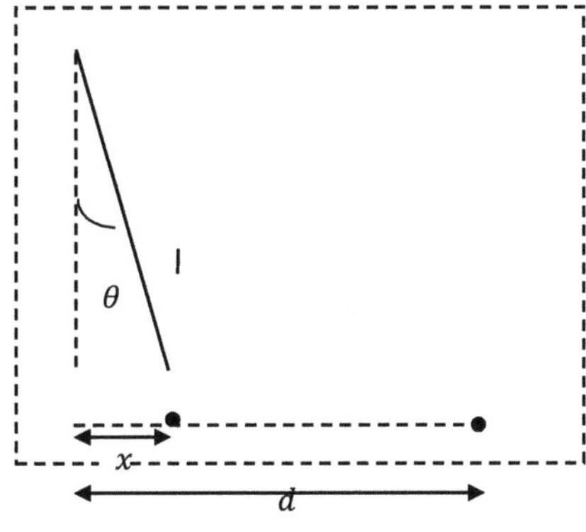

Figure 14

1. Calculate the angle θ of inclination of the pendulum.
2. Calculate the electrostatic force exerted on the ball.

Exercise 3 solution:

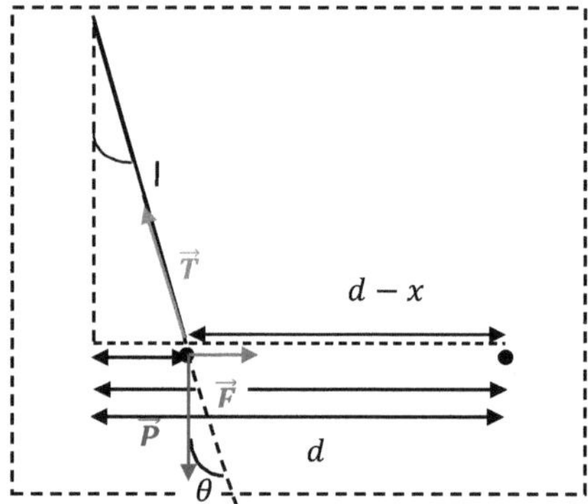

Figure 14

1. The angle θ of inclination of the pendulum

The electrical forces are attractive. Using the small angle approximation, we can write :

$$\frac{x}{l} = sin\theta = tan\theta = \frac{F}{m.g} \text{ Or } \quad F = -k\frac{Q_1Q_2}{(d-x)^2}$$

So :

$$\frac{x}{l} = -k\frac{Q_1Q_2}{(d-x)^2}$$

$$\Rightarrow (d-x)^2 = -k\,l\frac{Q_1Q_2}{m\,g}\frac{1}{x}$$

$$\Rightarrow x^3 - 2\,d\,x^2 + k\,l\frac{Q_1Q_2}{m\,g}\frac{1}{x} = 0$$

Hence :

$$x^3 - 8.10^{-2}x^2 + 16.10^{-4}x - 9.10^{-6} = 0$$

The equation can be simplified by setting$x = y.10^{-2}$ (working in cm).

The result is :

$$y^3 - 8\,y^2 + 16\,y - 9 = 0$$

Note that y = 1 is a solution. The equation then becomes :

$$(y - 1)(y^2 - 7\,y + 9) = 0$$

The two remaining solutions are$(\mathrm{y}^2 - 7\,\mathrm{y} + 9) = 0$. We find :

$$y = \frac{7 \pm \sqrt{13}}{2}$$

These three solutions correspond to x = 1 cm, x = 1.7 cm or x = 5.30 cm. The last solution corresponds to x > d, which contradicts the hypothesis of the exercise. What's more,$\theta = arcsin\frac{x}{l} = \arcsin 0{,}53° = 32°$ does not satisfy the approximation . $\sin\theta \simeq \theta$

The first two solutions are acceptable and correspond to differentθ while still satisfying the approximation$\sin\theta \simeq \theta$. We choose x = 1cm because it is the best approximation ($\theta = \arcsin 0{,}1 = 5{,}7°$is the smallest).

2. Electrostatic force

We write :

$$F = -k\frac{Q_1 Q_2}{(d-x)^2} = \frac{x}{l}\, m\, g = 10^{-2}\ \text{N}$$

Exercise 4:

Let four point charges be placed at the vertices of a square with side a = 2cm and centre 0 as shown in Figure 15.

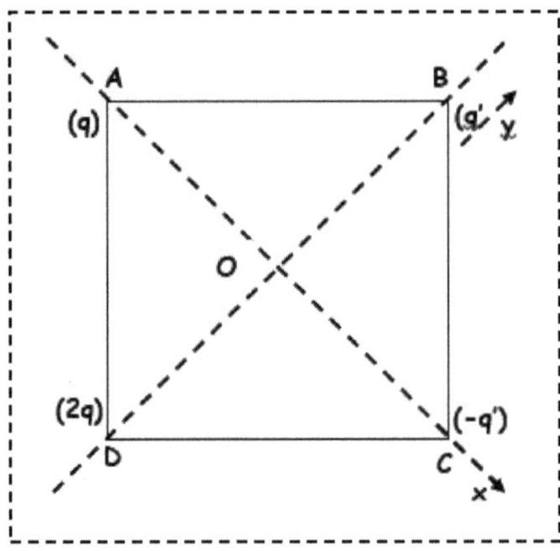

Figure 15

1. Knowing that the expression for the resulting field at point O is $\overrightarrow{E_o} = 5\vec{\imath} + 6\vec{\jmath}$ (v/m), determine the values of q and q'.
2. Determine the potential at point O.

3. A charge q" = 2 µC is placed at point O. Deduce its potential energy and the force exerted on it.

Exercise 4 solution:

1. The values of q and q'

Given that four charges are in the vicinity of the point O, the resulting electric field $\overrightarrow{E_O}$ is the sum of the electric fields created by each charge. Consequently, we write :

$$\overrightarrow{E_O} = \overrightarrow{E_{A/O}} + \overrightarrow{E_{B/O}} + \overrightarrow{E_{C/O}} + \overrightarrow{E_{D/O}} \quad (1)$$

Note that :

$$OA = OB = OC = OD = \frac{a}{\sqrt{2}}$$

With :

$$\overrightarrow{E_{A/O}} = 2k\frac{q}{a^2}\vec{\imath}, \overrightarrow{E_{B/O}} = -2k\frac{q'}{a^2}\vec{\jmath}, \overrightarrow{E_{C/O}} = 2k\frac{q'}{a^2}\vec{\imath},$$

$$\overrightarrow{E_{A/O}} = 4k\frac{q}{a^2}\vec{\jmath}$$

Then expression (1) becomes :

$$\overrightarrow{E_O} = \frac{2k}{a^2}(q+q')\vec{\imath} + \frac{2k}{a^2}(2q-q')\vec{\jmath} \quad (2)$$

Based on the data for the year, the value of $\overrightarrow{E_o}$ is :

$$\overrightarrow{E_o} = 5\vec{\imath} + 6\vec{\jmath} \text{ (v/m) (3)}$$

Then expressions (2) and (3) are equal. Consequently, we write

:

$$\begin{cases} \frac{2k}{a^2}(q+q') = 5 \\ \frac{2k}{a^2}(2q-q') = 6 \end{cases}$$

$$\Rightarrow \begin{cases} (q+q') = \frac{10}{9}10^{-13} \\ (2q-q') = \frac{12}{9}10^{-13} \end{cases}$$

$$\Rightarrow \begin{cases} \boldsymbol{q = 0,81\ 10^{-13}\ (C)} \\ \boldsymbol{q' = 0,29\ 10^{-13}\ (C)} \end{cases}$$

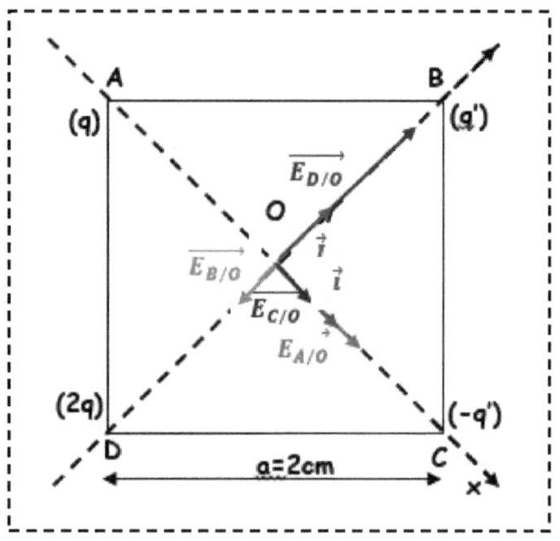

Figure 15

The potential V_0 at O

The resulting electric potential V_0 at the point O is the sum of the potentials , ,$V_A V_B V_C$ and V_D created respectively by the charges q , q', -q' and 2q. Consequently, we write :

$$V_O = V_A + V_B + V_C + V_D$$

With :

$$V_A = k\frac{q}{a}\sqrt{2}, ,V_B = k\frac{q'}{a}\sqrt{2} V_C = -k\frac{q'}{a}\sqrt{2} \text{ and } ,V_D = k\frac{q}{a}2\sqrt{2}$$

This gives :

$$\boldsymbol{V_O = k\frac{q}{a}3\sqrt{2} = 3,66\ 10^{-2}\ \mathrm{V}}$$

2. **A charge q''=2 μC is placed at the point O . Deduce its potential energy and the force exerted on it.**

We write :

$$\begin{cases} E_P = q''V_O = 7{,}33\ 10^{-8}\ J \\ \overrightarrow{F_O} = q''\overrightarrow{E_O} = 10^{-5}\ (\vec{\imath} + 1{,}2\,\vec{\jmath})\ N \end{cases}$$

Exercise 5:

Consider the system of point charges shown in Figure 16. The charges q and -q are placed at coordinates (0, -a) and (0, a) respectively, and the charge Q (positive) has coordinates (0, y) such that y> 0.

1. Give the expression of the force exerted on the load Q placed at point M, then calculate it and draw it to scale; 1cm = 20 N.

2. Find the y_0 ordinate so that the charge -q is in an equilibrium position.

Data: a=1cm, y=5cm. Q=2.10^{-6} C, q=3.2. C.10^{-6}

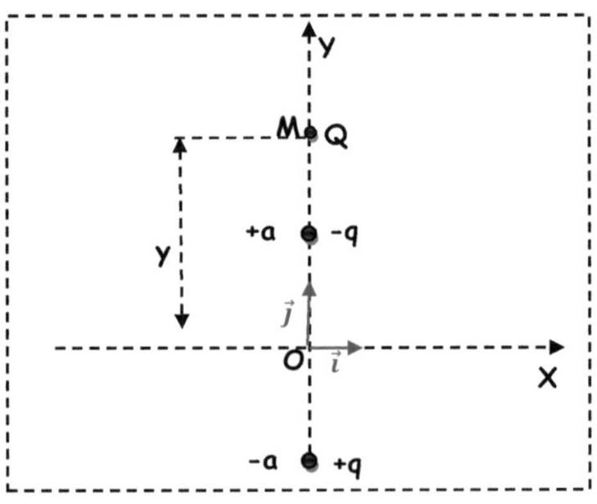

Figure 16

Exercise 5 solution:

1. Determination and representation of $\overrightarrow{F_M}$:

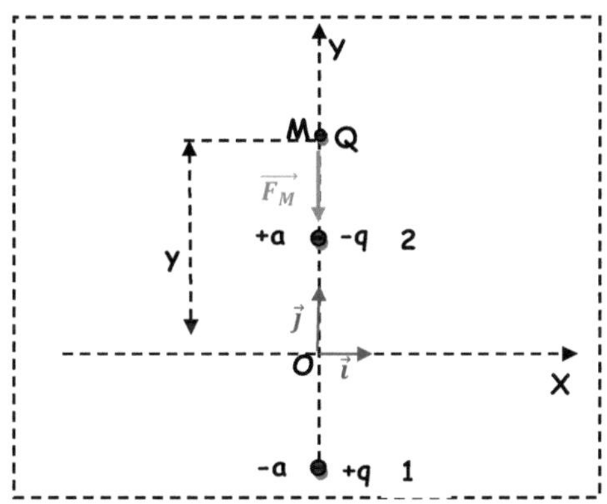

Figure 17

The resulting electric force $\overrightarrow{F_M}$ is the sum of the electric forces created by each charge at point M. Consequently, we write :

$$\overrightarrow{F_M} = \overrightarrow{F_{1/M}} + \overrightarrow{F_{2/M}}$$

With :

$$\overrightarrow{F_{1/M}} = \frac{k\,(-q)\,Q}{(y-a)^2}\vec{j} \text{ and } \overrightarrow{F_{2/M}} = \frac{k\,(+q)\,Q}{(y+a)^2}\vec{j}$$

This gives :

$$\overrightarrow{F_M} = \overrightarrow{F_{1/M}} + \overrightarrow{F_{2/M}} = k\,q\,Q\left(\frac{-1}{(y-a)^2} + \frac{1}{(y+a)^2}\right)\vec{j}$$

$$\boldsymbol{\overrightarrow{F_M}\,(y = 5cm) = -20\vec{j}\,(N)}$$

The force$\overrightarrow{F_M}$ is shown in Figure 4 to scale: 1cm→ 20 N

2. Determining the equilibrium position :

The charge -q is in equilibrium at position y_0 ⇨ $\vec{F}(y_0) = \vec{0}$

We write :

$$\vec{F}(y_0) = \overrightarrow{F_{M/1}} + \overrightarrow{F_{2/1}}$$

With :

$\overrightarrow{F_{M/1}} = \frac{k\,(-q)\,Q}{(y-y_0)^2}(-\vec{j})$ and $\overrightarrow{F_{2/1}} = \frac{k\,(-q)\,(+q)}{(y_0+a)^2}\vec{j}$

This gives :

$$\vec{F}(y_0) = k\,q\left(\frac{Q}{(y-y_0)^2} - \frac{q}{(y_0+a)^2}\right)\vec{j} = \vec{0}$$

$$\Rightarrow q\,(y-y_0)^2 = Q\,(y_0+a)^2$$

$$\Rightarrow \begin{cases} y_{01} = 2{,}69\ cm\ (Retenue) \\ y_{02} = 15\ cm\ (Non\ retenue) \end{cases}$$

$$\Rightarrow \boldsymbol{y_0 = y_{02} = 15\ cm}$$

Exercise 6:

A circular loop with centre O and radius R carries a uniformly distributed positive charge Q.

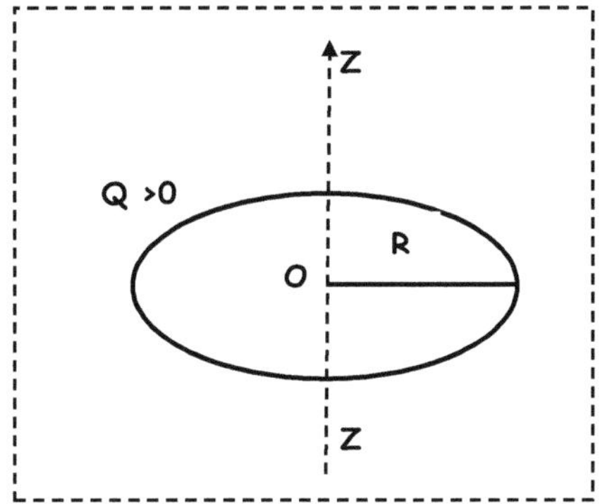

Figure 18

1. Determine the expression for the electric potential along the axis z'Oz perpendicular to the plane of the loop (Figure 17).
2. Give the expression for the electric field, using :
   a. Direct calculation.
   b. The expression of potential.

Exercise 6 solution:

1. The expression for the electric potential along the axis Z'OZ perpendicular to the plane of the loop

According to the data in the exercise, the loop carries a load$Q > 0$ distributed uniformly over the length of the loop. We write :

$$Q=\lambda\,.l=\lambda\,.\,2\pi R$$

By breaking down the length l of the loop into several elements dl each of which carries the charge dq , we write :

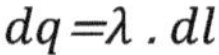

$$dq=\lambda\,.\,dl$$

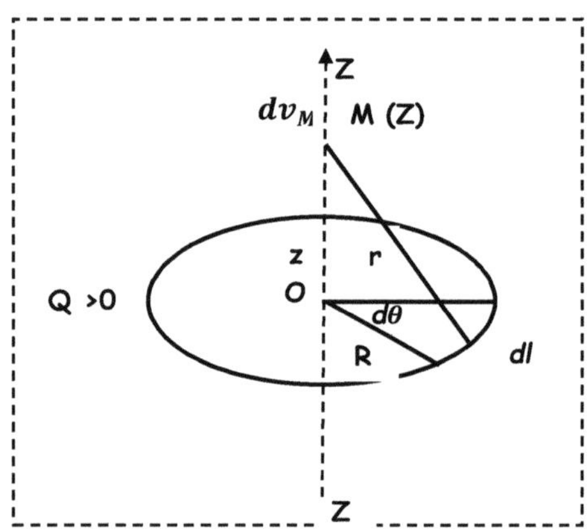

Figure 19

The element of charge dq creates in M an element of potential dv_m with :

$$dV_m = k\frac{dq}{r}$$

Note that :

$$\begin{cases} r = \sqrt{z^2 + R^2} \\ \quad et \\ dl = R.d\theta \end{cases}$$

So we write :

$$\partial V_M = k\frac{\lambda .R.d\theta}{\sqrt{Z^2+R^2}}$$

$$\Rightarrow \quad V_M = \int_l \ dV_M = \int_0^{2\pi} k \ \frac{\lambda \ R \ d\theta}{\sqrt{Z^2+R^2}}$$

$$\Rightarrow \quad V_M = k \ \frac{\lambda \ R}{\sqrt{Z^2+R^2}}\int_0^{2\pi} \partial\theta$$

$$\Rightarrow \quad \boldsymbol{V_M = K\frac{Q}{\sqrt{Z^2+R^2}}}$$

2. The expression for the electric field, using

a. The expression of potential

We have :

$$\vec{E}_M = -\overrightarrow{Grad}\ V_M$$

With :

$$\begin{cases} \vec{E}_M = E_X\vec{i} + E_Y\vec{j} + E_Z\vec{k} \\ \text{et} \\ \overrightarrow{Grad}\ V_M = \dfrac{\partial V_M}{\partial_X}\vec{i} + \dfrac{\partial V_M}{\partial_Y}\vec{j} + \dfrac{\partial V_M}{\partial_Z}\vec{k} \end{cases}$$

SinceV_M depends only on z, then :

$$\overrightarrow{Grad}\ V_M = \frac{\partial V_M}{\partial_Z}\vec{k}$$

So :

$$\vec{E}_M = -\frac{\partial V_M}{\partial_Z}\vec{k}$$

$$\Rightarrow \vec{E}_M = -\frac{\partial}{\partial_Z}\ k\frac{Q}{\sqrt{Z^2+R^2}}\vec{k}$$

$$\Rightarrow \quad \vec{E}_M = k\frac{Q\,Z}{\sqrt{Z^2+R^2}^{\,3/2}}\vec{k}$$

b. Expression of potential

The element of lengthdl which carries the element of chargedq creates at the point$M$ an element of electric field$d\vec{E}_M$, with :

$$d\vec{E}_M = k\frac{dq}{r^2}\vec{u}$$

For reasons of symmetry, the electric field $\vec{E}_M$ created by the whole loop is along the axis (OZ), so :

$$d\vec{E}_{m\,(Z)} = dE_M . \cos\alpha$$

$$E_{M\,(Z)} = \int dE_M . \cos\alpha$$

With :

$$\begin{cases} \cos\alpha = \dfrac{Z}{r} = \dfrac{Z}{\sqrt{Z^2+R^2}} \\ \quad et \\ dl = R.d\theta \end{cases}$$

So

$$E_{M\,(Z)} = \int_0^{2\pi} K\frac{\lambda\,R\,d\theta}{Z^2+R^2}.\frac{Z}{\sqrt{Z^2+R^2}} = K\frac{\lambda\,R\,Z}{\sqrt{Z^2+R^2}^{\,3/2}}\int_0^{2\pi} d\theta$$

$$\Rightarrow E_{M\,(Z)} = k\,\frac{\lambda.2\pi R\,Z}{\sqrt{Z^2+R^2}^{\,3/2}}$$

$$\Rightarrow \; E_{M\,(Z)} = k\,\frac{Q\,Z}{\sqrt{Z^2+R^2}^{\,3/2}}$$

$$\Rightarrow \qquad \vec{E}_{M\,(Z)} = E_M \vec{k}$$

$$\Rightarrow \; \vec{E}_{M\,(Z)} = k\frac{Q\,Z}{\sqrt{Z^2+R^2}^{\,3/2}}$$

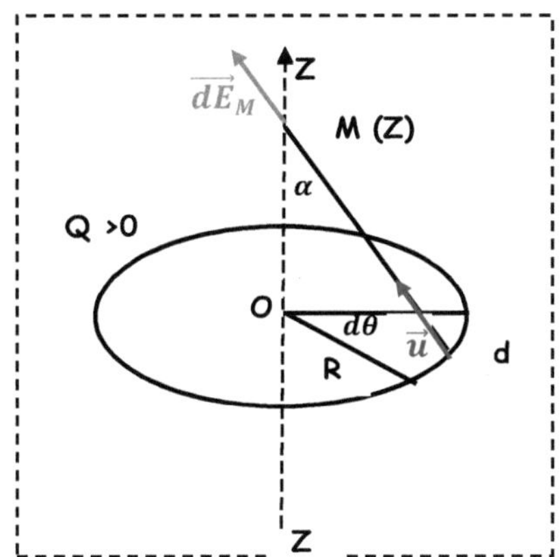

Figure 17

Exercise 7

Is it practical to apply Gauss's theorem to determine the electric field in the following cases? Justify this.

1. An irregularly shaped, uniformly charged body.
2. A wire or cylinder of finite length uniformly loaded.
3. An infinite wire, an infinite plane, a sphere or a cylinder of infinite length charged with a variable charge density.
4. A uniformly loaded wire or cylinder of infinite length.

Exercise 7 solution:

Compared with the direct calculation of the electric field$\vec{E}$, Gauss's theorem can be applied if symmetry considerations are favourable, in particular the norm of the field$\vec{E}$ is the same at all points on the Gauss surface. For this reason, it is not practical to apply Gauss's theorem to determine the electric field in cases 1 and 2. However, it can be applied in cases 3 and 4.

Exercise 8

Let a straight wire of infinite length be uniformly loaded with a density $\lambda > 0$.

1. Calculate the fields and the potential at any point in space. We give V(1)= 0

2. Draw the field lines and equipotentials. (The Gauss surface for a straight wire is a closed cylinder with the wire at the centre).

Exercise 8 solution:

1. The expression for the electric field $\vec{E}\,(r)$ and the potential V(r):

a. The expression for the electric field : $\vec{E}\,(r)$

For reasons of symmetry, the electric field created by the infinite wire is radial, and the Gaussian surface in this case is a closed cylinder of radius r and height h where the wire is at the centre of the cylinder (Figure 18).

Following Gauss's theorem, we write :

$$\emptyset = \oiint \vec{E}\,.\overrightarrow{dS} = \frac{\sum_i Q_{int}}{\varepsilon_0}$$

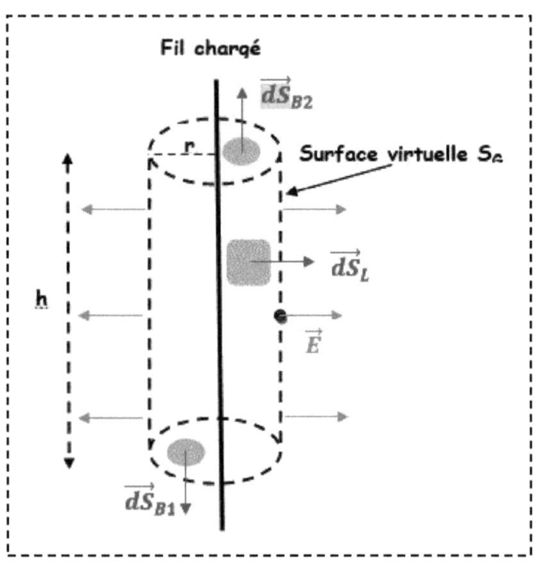

Figure 20

The total flux through the Gaussian surface:

$$\emptyset_{S_G} = \emptyset_{S_{B1}} + \emptyset_{S_{B2}} + \emptyset_{S_L}$$

$$\emptyset = \oiint \vec{E} \cdot \overrightarrow{dS} = \iint_{S_{B1}} \vec{E} \cdot \overrightarrow{dS}_{B1} + \iint_{S_{B2}} \vec{E} \cdot \overrightarrow{dS}_{B2} + \iint_{S_L} \vec{E} \cdot \overrightarrow{dS}_L$$

⇨

As($\vec{E} \perp \overrightarrow{dS}_{B1}$) and($\vec{E} \perp \overrightarrow{dS}_{B2}$) the field is radial:

$$\iint_{S_{B1}} \vec{E}.\overrightarrow{dS_{B1}} = \iint_{S_{B2}} \vec{E}.\overrightarrow{dS_{B2}} = 0$$

And as($\vec{E}$ // $\overrightarrow{dS_L}$) and in the same sense, then :

$$\iint_{S_L} \vec{E}.\overrightarrow{dS_L} = E(r).dS_L$$

This gives us :

$$\emptyset = \oiint \vec{E}\,.\overrightarrow{dS} = E(r).dS_L$$

Hence :

$$\emptyset = \boldsymbol{E(r)\,.2\pi r h} \quad \textit{(1)}$$

Inside the Gauss surface the charge is carried by the wire (uniform distribution):

$$\boldsymbol{\sum_i Q_{int} = \lambda\,.\,l = \lambda\,.\,h} \quad \textit{(2)}$$

From equations (1) and (2) :

$$E(r)\,.2\pi.r.h = \frac{\lambda.h}{\varepsilon_0}$$

Then the field created by the wire at any point in space is :

$$\boldsymbol{\vec{E} = \frac{\lambda}{2\pi\varepsilon_0 r}\,\vec{u}_r}$$

b. The expression for electrical potential

We know that :

$$\vec{E} = -\overrightarrow{grad}\, V$$

With :

$$\overrightarrow{grad}\, V = \frac{\partial V}{\partial_r}\vec{u}_r + \frac{1}{r}\frac{\partial V}{\partial_\theta}\vec{u}_\theta + \frac{\partial V}{\partial_Z}\vec{k} = \frac{\partial V}{\partial_r}\vec{u}_r$$

Since$V$ depends only on r, then

$$\overrightarrow{grad}\, V = \frac{\partial V}{\partial_r}\vec{u}_r$$

So..:

$$dv = -E.dr$$

$$\Rightarrow \int dv = -\int E.dr$$

$$\Rightarrow V(r) = -\frac{\lambda}{2\pi\varepsilon_0}\int \frac{1}{r}\, d_r$$

$$\Rightarrow \boldsymbol{V(r) = -\frac{\lambda}{2\pi\varepsilon_0}.\, Ln\, r + C}$$

If at (r = 1) the electric potential V is zero (V(1) = 0), then :

$$V(1) = -\frac{\lambda}{2\pi\varepsilon_0}.\, Ln\, 1 + C$$

$$\Rightarrow C = 0$$

Hence the potential created by a straight wire of infinite length is

$$V(r) = -\frac{\lambda}{2\,\pi\,\varepsilon_0}.\,Ln\,r$$

2. Field lines and equipotentials

Field lines: radial lines coming out of the wire, perpendicular to equipotential surfaces.

Equipotential surfaces: coaxial cylinders whose axes coincide with the wire (see figure 19).

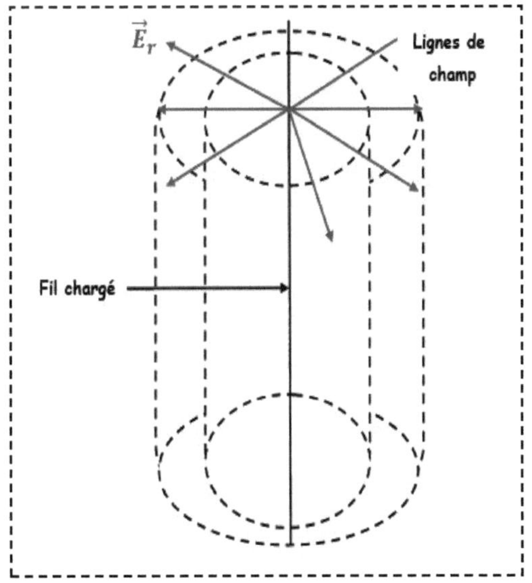

Figure 21

Exercise 9

A sphere of radius R carries a positive charge whose density depends only on the distance from its centre, such that$\rho = \rho_0 \left(1 - \frac{r}{R}\right)$ orρ_0 is constant.

Determine the electric field$\vec{E}$ throughout space, and what distancer_m this field is at its maximum.

Exercise 9 solution:

1. The expression of the electric field$\vec{E}\,(r)$ throughout space:

The sphere is charged in volume with a variable volume density.

For reasons of symmetry, the field created by this charge distribution is radial and the Gaussian surface is a sphere with radius r and the same centre as the real sphere.

Applying Gauss's theorem, we write :

$$\emptyset = \oiint \vec{E}\,.\overrightarrow{dS} = \frac{\sum_i Q_{int}}{\varepsilon_0}$$

There are two zones:

- **Zone I, r < R**

The electric field is radial and centrifugal, because the charge is positive

$$\rho = \rho_0 \left(1 - \frac{r}{R}\right)$$

So..:

$$\emptyset = \iint_{S_G} \vec{E}.\overrightarrow{dS}$$

$$\Rightarrow \emptyset = E.S_G$$

$$\emptyset = \boldsymbol{E.4\pi.r^2} \quad \textit{(1)}$$

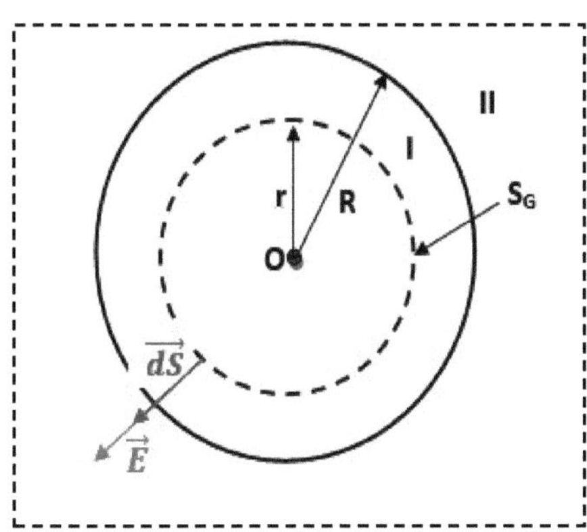

Figure 22

The load distribution is volumetric:

$$Q_{int} = \iiint_{S_G} \rho . dv$$

$$\Rightarrow \ Q_{int} = \iiint_{S_G} \rho_0 \left(1 - \frac{r}{R}\right) . dv \qquad (2)$$

Since the load distribution is radial :

$$dv = 4\pi . r^2 . dr$$

Then expression (2) becomes :

$$Q_{int} = \iiint_{S_G} \rho_0 \left(1 - \frac{r}{R}\right) . 4\pi . r^2 . dr$$

$$\Rightarrow \ Q_{int} = 4\pi . \rho_0 \left[\int_0^r r^2 . dr - \int_0^r \frac{r^3}{R} . dr \right]$$

Hence :

$$\boldsymbol{Q_{int} = 4\pi . \rho_0 \left[\frac{r^3}{3} - \frac{r^4}{4R} \right]} \qquad (3)$$

Following Gauss's theorem, we write :

$$\emptyset = \oiint \vec{E} . \overrightarrow{dS} = \frac{\sum_i Q_{int}}{\varepsilon_0}$$

From equations (1) and (3) :

$$E . 4\pi . r^2 = \frac{4\pi . \rho_0}{\varepsilon_0} \left[\frac{r^3}{3} - \frac{r^4}{4R} \right]$$

Then the field created by the sphere in zone I (r < R) is :

$$E = \frac{\rho_0}{\varepsilon_0}\left[\frac{r}{3} - \frac{r^2}{4R}\right]$$

- **Zone II, r≥ R**

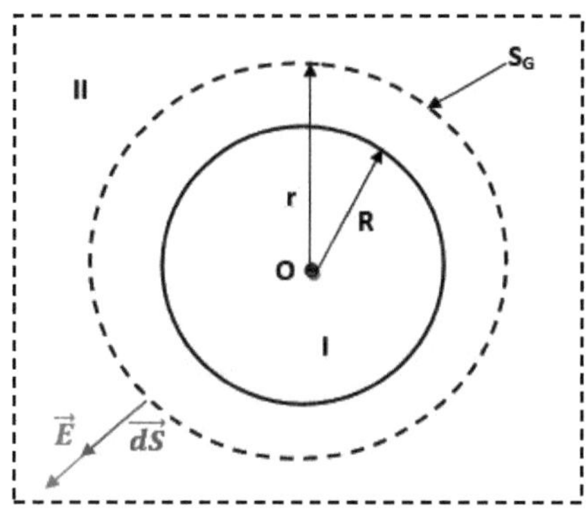

Figure 23

In the same way, to calculate the flow, we write :

$$\emptyset = \iint_{S_G} \vec{E}.\overrightarrow{dS}$$

$$\Rightarrow \emptyset = E.S_G$$

$$\emptyset = E.4\pi.r^2 \quad (4)$$

The charge distribution is volumetric (the charge is distributed in the volume of the real sphere (r varies from 0 to R) :

$$Q_{int} = \int_0^R \rho_0 \left(1 - \frac{r}{R}\right). 4\pi. r^2. dr$$

$$\Rightarrow Q_{int} = 4\pi. \rho_0 \left[\int_0^R r^2. dr - \int_0^R \frac{r^3}{R}. dr\right]$$

$$\Rightarrow Q_{int} = 4\pi. \rho_0 \left[\frac{R^3}{3} - \frac{R^4}{4R}\right]$$

Hence:

$$\boldsymbol{Q_{int} = \pi. \rho_0 \frac{R^3}{3}} \qquad \textbf{(5)}$$

Following Gauss's theorem, we write :

$$\emptyset = \oiint \vec{E}\,.\overrightarrow{dS} = \frac{\sum_i Q_{int}}{\varepsilon_0}$$

From equations (4) and (5) :

$$E. 4\pi. r^2 = \frac{\pi. \rho_0. R^3}{3. \varepsilon_0}$$

Then the field created by the sphere in zone II (r≥ R) is :

$$\boldsymbol{E = \frac{\rho_0. R^3}{12. \varepsilon_0}\left[\frac{1}{r^2}\right]}$$

Then the field created by the sphere at any point in space is given by :

$$\vec{E}(r) = \begin{cases} \frac{\rho_0}{\varepsilon_0}\left[\frac{r}{3} - \frac{r^2}{4R}\right].\vec{u}_r, r < R \\ \frac{\rho_0.R^3}{12.\varepsilon_0}\left[\frac{1}{r^2}\right].\vec{u}_r, r \geq R \end{cases}$$

2. Calculate the distance r_m where the field is strongest:

The maximum of the electric field is where its derivative is cancelled out by changing sign:

	Zone I						Zone II
r	0		$\frac{2}{3}R$		R		∞
$\frac{dE}{dr}$		+	0	-			-
E (r)	0	↗	$\frac{\rho_0.R}{12.\varepsilon_0}$	↘	$\frac{\rho_0.R}{12.\varepsilon_0}$	→	0

According to this table of variation the field is maximal for :

$$r_m = \frac{2}{3}R$$

References

[1] Dr. Afifa Yedjour, Dr. H. Aouabdi-Settaouti, Electricity, University of Science and Technology of Oran, Mohamed Boudiaf USTO-MB. Algeria, - 2016/2017.

[2] Mr Abdeladim Mustapha, Polycopié de Cours Physique 2, Université des sciences et technologie d'Oran, Mohamed Boudiaf, Algérie, - 2015/2016.

[3] M. Berlin, J.P. Faroux and J. Renault, "Electromagnetisme 1, Electrostatique", Dunod, 1977.

[4] Dr.Samir Khene, Electricité Notion de Base, Electrostatique, Electrocinétique, Electromagnétisme, Rappels de cours et exercices corriges - OPU.

[5] E. Amzallag, J. Cipriani, J. Ben Naim and N. Piccioli "La physique du Fac, Electrostatique et Electrocinétique" 2nd Edition, Edi-Science, 2006.

[6] Pr. M. Chafik El Idrissi, Electricité cours exercices et problèmes corrigés, Faculté des sciences, Université Ibn Tofail. Kénitra.

[7] A. Fizazi, "Electricity and Magnetism", OPU, 2012.

[8] Séries de travaux dirigés " Electricité " et Examens, Université des sciences et technologie Abd El Hamid Ibn Badis, Mostaganem 2015-2020.

[9] Mahmoud Hachemane, Electricity exercises with solutions, Faculty of Physics, USTHB, 2014.

Printed by Books on Demand GmbH, Norderstedt / Germany